10X AI
AMPLIFIER

Unlock the Tools, Mindset, and Moves to Multiply Your Business

Scott Sullivan
Jessica Fontana-Eddowes

Contents

Introduction
Welcome to the 10X AI Amplifier

The Window Is Closing. Your Competitors Are Already Moving.

You didn't pick up this book by accident. You're here because something in your gut is whispering—or maybe screaming—that you could be doing more. Scaling faster. Leading smarter. Building something bigger than what you're currently seeing in front of you.

And deep down, you know that Artificial Intelligence might be the missing key.

But here's what you might not realize: **while you're wondering where to begin, thousands of business leaders have already begun.**

Right now, as you're reading this:

The marketing agency owner is producing 3x more content with the same team while working 20 fewer hours per week. **The consultant** transformed his proposal process from 12 hours to 2 hours—with a 40% higher close rate. **The e-commerce**

entrepreneur automated her customer service and inventory management so completely that she's expanding into new markets while her competitors drown in daily operations.

What they all have in common: They didn't wait for permission. They didn't wait for perfection. They started with one system, proved it worked, then built from there.

And they're pulling further ahead every day.

The Great AI Separation Is Happening Now

We're living through what will be remembered as the Great AI Separation of the 2020s. On one side: leaders who embrace AI as a strategic amplifier. On the other: those clinging to manual processes and outdated systems.

The gap is widening daily.

Every week you wait, your AI-empowered competitors:

- Process information faster and make decisions with better data
- Serve more customers with higher quality
- Scale their impact without scaling their stress
- Build systems that work while they sleep

Meanwhile, businesses stuck in "traditional" modes are working harder than ever just to maintain their current output.

This isn't a future scenario. It's your current reality.

Why We Wrote This Book (And Why You Need It Now)

We didn't write this book because AI is trendy. We wrote it because AI is *inevitable*—and most business owners are about to get left behind.

We're not here to give you hype. We're here to give you **leverage**. To show you how to go from being overwhelmed by AI to owning it as your most powerful amplifier.

This book was born out of real, gritty experience: **Scott**, a business builder who's mastered communication and growth psychology, and **Jessica**, a systems strategist who sees the straight lines others can't.

Together, we've:

- Led businesses through massive scaling transformations
- Bought and optimized millions in media spend
- Built comprehensive tech stacks and high-conversion funnels
- Led international teams across multiple countries
- Coached thousands on unlocking their full potential

Most importantly: We've guided dozens of businesses through successful AI transformations. We know where the dangerous paths are, which shortcuts actually work, and how to avoid the costly mistakes others have made.

Think of us as your AI Sherpas. We've already climbed this mountain. We know the route to the summit.

What You'll Gain: The Complete 10X AI Amplifier

This isn't another book about "the future of AI." This is your field manual for dominating in the AI era. Here's what you can expect as you move through these pages:

☑ **Clarity** — What AI really is, what it isn't, and what it can do for you right now
☑ **Confidence** — Learn to use AI tools without needing to be a tech wizard
☑ **Speed** — Reclaim 5–10 hours a week using plug-and-play automations

- ☑ **Simplicity** — Frameworks that cut through the noise and complexity
- ☑ **Scale** — Grow your message, business, and systems without losing your soul
- ☑ **Strategic Edge** — Build workflows and tech stacks that actually work together
- ☑ **Team Empowerment** — Bring your people along without triggering panic or resistance
- ☑ **Customer Intimacy** — Use AI to personalize at scale while deepening connection
- ☑ **Leadership Evolution** — Step into the version of you that can lead in this new era
- ☑ **Exponential Results** — Not just 10% better—but 10X. Consistently.

How This Book Works: Your Transformation Roadmap

The structure is simple but strategic. Each part builds on the last, guiding you through a natural evolution from overwhelm to exponential growth:

Part I: Rewiring the Mindset

We start where most skip: your brain. Your mindset is the biggest multiplier—or barrier—to AI-powered growth. We'll help you release fear, upgrade your perspective, and see new possibilities that your competitors are missing.

Part II: Strategic Foundations of Amplification

Now that you can see clearly, we help you build systematically. You'll get ready-to-use prompts, workflow templates, and a new model of business built for intelligent leverage.

Part III: Tools, Tech & Trust

We'll demystify tech stacks, break down tools by ease and ROI, and show you how to keep your voice and values intact while scaling. You'll also learn how to bring your team with you—without resistance.

Part IV: The Exponential Path Forward

Once you're in flow, we help you fly. Learn how to 10X your results, scale without burnout, and step fully into **augmented leadership**—where AI doesn't replace your intelligence, but extends it.

You won't just learn to "use AI." You'll learn how to **think with it**. To treat it as a cognitive amplifier—a strategic appendage that processes faster, sees wider, and acts with precision.

This isn't about becoming less human. It's about becoming **more capable**—gaining a digital extension of yourself that makes you more focused, effective, and future-ready.

Your Sneaky Competitors Are Already Moving

Most people are watching AI from a distance, paralyzed by complexity or stuck in analysis paralysis.

But your sneakiest competitors? They're already using it to silently outpace you:

- Sending 10X more personalized emails
- Publishing high-quality content consistently
- Converting leads while they sleep
- Making data-informed decisions in minutes, not months

This book helps you become the competitor others can't keep up with.

You don't need to code. You don't need to hire a Silicon Valley engineer. You need to understand the principles, shift your perspective, and apply the practices we outline.

With expert guidance to avoid the costly mistakes that trip up most first-time AI adopters.

You Don't Have to Navigate This Alone

Here's what we've learned from working with hundreds of business owners: **The leaders who transform fastest have experienced guides.**

The ones who try to figure it out alone? They get lost in complexity, waste months on systems that don't integrate, or give up when they hit the inevitable obstacles.

The ones who work with AI Sherpas reach the summit faster, safer, and with more confidence.

That's why throughout this book, you'll find opportunities to connect with our team of Fractional AI Officers. We're not just authors—we're practitioners who build these systems daily for businesses like yours.

Starting with our Strategic AI Readiness Diagnostic—a comprehensive assessment that shows you exactly where you are, where the biggest opportunities lie, and your personalized roadmap to exponential growth.

Your Limited Window of Opportunity

Here's what most business books won't tell you: **You have a limited window of competitive advantage.**

Right now, AI adoption is still in early stages. Leaders who act in the next 6-12 months will establish advantages that become nearly impossible for competitors to close.

Every month you wait:
- Your competitors get further ahead
- Customer expectations rise based on AI-powered experiences
- The learning curve gets steeper as the complexity gap widens

Every month you act:

- You build capabilities that compound over time
- You develop AI fluency while it's still a competitive advantage
- You position yourself as a leader, not a follower

You Are the Amplifier

The greatest myth is that AI will replace humans. It won't.

But it **will** replace humans who refuse to adapt.

You don't have to be one of them.

This is your invitation to become someone who:

- Scales smarter, not harder
- Leads with intelligence, not just instinct
- Sees opportunity where others see overwhelm
- Amplifies their purpose through intelligent systems

The choice is yours. But the opportunity won't wait.

After you finish this book, you'll face a decision:

Option 1: Think "that was interesting" and go back to business as usual. Watch from the sidelines as AI-powered competitors pull ahead.

Option 2: Take immediate action. Start with our Strategic AI Readiness Diagnostic. Get expert guidance. Build momentum through intelligent experimentation.

The leaders choosing Option 2 are already transforming their businesses.

The ones choosing Option 1 will spend the next few years trying to catch up.

Which will you choose?

Let's get started.

Chapter 1
Your First New Playbook –
Seeing Differently

Harnessing AI to Unlock the Growth That's Already Around You

You're not behind because you're lazy. You're behind because the game changed—and no one handed you the new playbook.

If you're like most driven leaders, you're running hard but not gaining ground. You sense your business is capable of more, but the path forward feels foggy. That's not because the opportunity isn't there. It's because you're looking at it through the wrong lens.

This chapter will hand you a new one.

The shift we're talking about isn't theoretical. It's already happening. Every week, entrepreneurs use AI to multiply their reach, automate their workflows, and accelerate decision-making. They're doing in hours what used to take days. They're reaching thousands without burning out. And they're building companies with fewer people and more profit.

They didn't get smarter. They started seeing differently.

The Old Lens vs. the New Lens

Old thinking says: work harder, hire more people, do more with less.

New thinking says: integrate intelligence, automate the repeatable, amplify what only you can do.

Most business owners get stuck trying to solve tomorrow's problems with yesterday's solutions. They stay buried in task lists and to-dos, hoping clarity will come once things "slow down." But that slowdown never comes. Why? Because they never shift their frame of reference.

Let's change that—right now.

Why You Can't Afford to Wait

AI isn't coming. It's here. And it's not just for the "tech guys." It's for entrepreneurs who want time back, leaders who want clarity, and creators who want to scale without losing their soul.

You don't need a computer science degree. You need a decision. A willingness to explore what's possible when you pair your vision with the right form of intelligence.

Here's what you gain when you embrace AI as an amplifier:
- You stop reacting and start anticipating.
- You stop grinding and start scaling.
- You stop drowning in busywork and start leading with strategy.

From Chaos to Clarity—
How K to Z Used AI to Reclaim Time and Reignite Growth

Brandon Barton, owner of K to Z Interiors & Outdoor Living, wasn't afraid of hard work. He'd grown the business from a small family side project into a multi-crew operation serving Louisiana with premium indoor and outdoor window treatments. But in early 2024, he found himself running harder than ever—and getting nowhere.

His team was overwhelmed. Leads came in through web forms, phone calls, and builder uploads—but there was no system to catch them. Admins like Kelsey were manually tracking builder POs, chasing missed calls, compiling spreadsheets, and emailing reminders when time allowed. Sales data was fragmented. Installers reported by text. And hundreds of thousands in quotes sat untouched each month because there was no time—or process—for structured follow-up.

The opportunity wasn't missing. It was leaking through the cracks.

When Brandon and his team partnered with Samurai Partners and integrated an AI-powered automation system and operational, everything changed. AI didn't replace his people. It amplified them.

- Incoming leads were routed, tagged, and responded to automatically.
- Smart follow-ups re-engaged customers who had gone dark.
- Installers submitted mobile forms with photos and updates, eliminating text thread chaos.
- Sales dashboards finally made quote-to-close tracking visible.

Kelsey got back 30% of her day. Brandon gained visibility into what was working. They no longer guessed where to spend effort—they knew.

With AI doing the heavy lifting, the team could focus on what they do best: educating customers, providing exceptional service, and closing with confidence.

This wasn't a theoretical shift. It was operational clarity—built in weeks. And it reminded them of something powerful: the growth they'd been chasing wasn't ahead of them. It had been there all along, buried under busywork.

They just had to see differently.

You Don't Have a Tool Problem — You Have a Vision Problem

Most books jump straight to the tools: "Use this app, try that tool." But tactics without clarity create chaos. If you don't know where you're going, every tool becomes another distraction.

The goal of this book is different. We'll get to the tactics—but first, we help you see. Not just with your eyes, but with your strategy. We help you shift from *How do I keep up?* to *How do I lead?*

Because if you see AI as just another to-do list item, you'll burn out. But if you see it as a **strategic amplifier**, you'll finally feel the momentum that's been missing.

Case in Point: The Hidden Opportunity Trap

Let's say you spend an hour every day writing emails. That's five hours a week, 260 hours a year. Now imagine an AI assistant that drafts those emails for you in minutes, tailored to your voice and strategy. You still review and refine, but you've just won back an entire week and a half of your life every year.

Now multiply that across every repetitive function in your business—client follow-up, data entry, content ideation, scheduling. AI doesn't just shave off minutes. It unlocks entire **layers** of time and energy you didn't know you were losing.

Let's Be Clear: AI Won't Replace You. But It Will Replace Something.

AI won't replace heart, connection, or creativity. But it will replace:

- Wasted hours
- Clunky processes
- Missed opportunities
- Decisions made in the dark

It's already replacing businesses that refuse to adapt. But for those willing to evolve, it's handing out unfair advantages daily.

Which one do you want?

The Real Starting Point: A New Frame of Reference

You won't scale by doing more. You'll scale by thinking differently.

The biggest leap isn't learning how to prompt ChatGPT or automate a workflow. It's training your mind to look for leverage in every corner of your business—and to partner with the intelligence that already sees it.

From this point forward, don't ask "Can I do this faster?"

Ask "Should I be doing this at all?"

Because AI isn't here to do more of the same. It's here to help you do **less of what drains you**, and **more of what moves the needle**.

Your Move

In the next chapter, we'll walk through the exact mindset upgrades that create exponential gains. But first, ask yourself:

- Where am I still trying to scale through effort instead of intelligence?
- What would it look like if AI didn't just assist me — it **amplified** me?
- If I could delegate one category of work to an intelligent system, what would it be?

Write those answers down. They're your first signals of where the shift begins.

Welcome to the new lens.

Welcome to the 10X AI Amplifier.

Chapter 2
The Mindset That Multiplies

Why What You Believe About AI Matters More Than What You Know

The fastest way to 10X your business isn't through tools.

It's through thinking.

Yes, we'll talk about prompts, platforms, and automations. But none of that matters if your mental operating system is outdated. Mindset is either the multiplier or the limiter of everything AI can do for you.

This chapter is about removing the mental governor from your growth engine. Because no matter how smart your strategy or how powerful your tech stack, you can't scale what your mind won't allow.

The Three Mindset Blocks That Kill AI Momentum

Before we can accelerate, we need to clear the road. These are the three most common mindset traps we see in business owners and teams trying to scale with AI:

1. The "I'm Not Techy" Identity

This is one of the most dangerous labels you can give yourself. It's not humility—it's self-sabotage. You don't need to *be techy* to leverage AI. You need to be willing to learn, test, and iterate. That's it.

2. The "That's Not Me" Syndrome

You believe your business is "too human," "too relational," or "too personal" to use automation. But automation isn't about replacing intimacy—it's about **protecting your time** so you can *create more of it*.

3. The "I'll Just Wait and See" Trap

AI is moving exponentially. The longer you wait, the wider the gap becomes. You don't need to master everything today, but you **do** need to begin. Momentum favors the early.

Reframing the Fear: AI as a Mirror, Not a Monster

Here's what most people miss: AI doesn't change who you are

—it **amplifies** who you already are.

If you're clear, strategic, and service-focused, AI will make you even more so. If you're disorganized, reactive, and chaotic, AI will multiply that too.

That's not a threat. That's a gift. Because it gives you a chance to pause, get aligned, and consciously design what you want AI to scale.

AI won't fix a broken model. But it will pour gasoline on one that works.

From Effort to Exponential

We've all been taught to grind. To hustle harder, longer, and louder. But that model is dead. AI introduces a new paradigm: **less friction, more flow**.

Think about it this way:

- Effort says, "How can I get this done faster?"
- Exponential asks, "How can I make sure I never do this again?"

Effort solves the surface. Exponential solves the system.

If you want to 10X, stop asking how to move faster on the hamster wheel. Start asking how to step off it altogether. Or, if AI can spin the wheel for you.

Build Your New Mental OS

Your brain is powerful—but it needs an upgrade. Here's the new internal software we'll be installing over the coming chapters:

- **From "How do I do this?" → "How can this be done without me?"**
- **From "What's the cost?" → "What's the opportunity cost if I don't?"**
- **From "I need to learn more first." → "I'll learn while executing."**
- **From "I need to do everything." → "I only do what moves the needle."**

This isn't ego. This is efficiency. You weren't built to be the bottleneck. You were built to lead.

What Happens When You Get This Right

When your mindset aligns with the exponential capabilities of AI:

- **You stop second-guessing and start testing.**
- **You stop hoarding decisions and start delegating to systems.**
- **You stop fearing change and start directing it.**

This mental pivot unlocks real results: more revenue, more peace, more capacity, and more time.

It also changes how you show up—as a visionary, not just an operator.

Questions to Move You Forward

- Where am I telling myself "I'm not ready" when I really mean "I'm afraid"?
- What do I believe about AI—and what might I need to unlearn?
- What part of my business am I still trying to control out of fear?

Write your answers. Sit with them. Then flip the fear into a frame of possibility.

In the chapter four, we'll give you your on-ramp

— the **First Five Moves** — to start integrating AI in a way that's low-risk, high-reward. You'll move from mindset into motion, with clarity and confidence.

But first, remember:

Your results won't grow faster than your beliefs allow.

Let's expand what's possible.

Chapter 3
Fear, Resistance
& the Human Side of Change

How to Emotionally Lead Yourself and Others Through the Shift

You can have the right strategy, the right tools, even the right timing—but if fear takes the wheel, you'll stall out before the first move.

AI isn't just a technological shift. It's an emotional one. And most books miss this. They tell you what to do, but not how to move through the invisible resistance that shows up when you try.

We take a different approach. We'll walk with you through the emotional side of this shift—so you don't just adopt AI, you activate everything around it.

Whether it's you, your team, or your clients—fear is part of the process. Resistance isn't a red flag. It's a sign that change is happening. And when you learn to lead through it, you don't just adopt AI—you amplify everything around it.

Resistance Is Natural—Until It's Not

Change always creates friction. But unacknowledged fear festers into paralysis.

What kind of fear shows up most?

- Fear of being replaced ("Will AI take my job?")
- Fear of irrelevance ("I'm too late to catch up.")
- Fear of overwhelm ("It's all too much to learn.")
- Fear of judgment ("What if I get it wrong?")
- Fear of losing the human touch ("My business is built on relationships.")

These aren't "technical issues." They're emotional defaults—and left unchecked, they quietly kill progress.

Fears in the Age of AI

Fear of Losing Human Touch
Concerns about AI affecting personal connections

Fear of Replacement
Concerns about AI taking over jobs

Fear of Judgment
Anxiety about making mistakes in AI adoption

Fear of Irrelevance
Feeling outdated in a rapidly changing world

Fear of Overwhelm
Feeling burdened by the amount of new information

Made with 🍃 Napkin

The Three Faces of Resistance

Resistance shows up differently in different people. Learn to spot it early.

1. The Disengager

They check out. They say "Sure, sounds good," but never follow through. Their resistance hides behind silence.

2. The Debater

They question everything. "Why should we trust AI?" "How do we know it won't backfire?" Their resistance hides behind logic.

3. The Doomsayer

They speak fear into the room. "This will never work." "This is dangerous." Their resistance hides behind urgency and drama.

Don't judge it. Understand it. Behind every reaction is a nervous system trying to stay safe.

⊛ Field Note: Leading Through Resistance—Kelsey's Story

When we first sat down with Kelsey, K to Z's Office Manager/Project Coordinator, her calendar was full and her hands were even fuller. She wasn't resistant because she didn't care—she was overwhelmed because she cared too much. Every missed follow-up, every builder PO she had to manually enter, every text thread with an installer—she was carrying all of it.

And then we said something that changed everything:

"This isn't about taking anything away from you. It's about freeing you to do the parts of the job only you can do."

Her whole posture changed.

Kelsey wasn't a Disengager or a Doomsayer. She was quietly debating whether this would mean more work disguised as "tech support." But when we reframed automation not as a threat, but as a promotion—something clicked.

She started naming things we could automate. She got excited about dashboards. She wanted to know how soon she could stop copying form fills into spreadsheets.

That's the shift. Not just getting buy-in—but building belief.

Reframing the Shift

Here's how to reposition AI for yourself and others:

From Replacement to Amplification: "AI won't replace you—it'll remove what's beneath you. So you can operate at your highest level."

From Threat to Opportunity: "Yes, AI will replace something. Let's make sure it replaces wasted time, not your role."

From Confusion to Clarity: "You don't need to know everything. You just need to start."

From Control to Collaboration: "You don't have to micromanage every step. AI becomes your second brain—so you can focus on what only you can do."

This speaks directly to team members like Kelsey who feel responsible for "holding it all together" and are afraid of losing oversight.

These reframes lower defenses. They signal safety, not stress.

Leading People Through the AI Curve

If you're leading a team, department, or community, here's your playbook of First 5 Moves to start with:

1. **Normalize the fear.** Say it out loud. "It's okay to feel uncertain. We're all adjusting."
2. **Clarify the why.** Tie AI to purpose. "We're not adopting AI to replace people. We're adopting it to better serve people."
3. **Start with wins.** Give people quick, visible victories. "Try this 5-minute prompt hack. Save an hour."
4. **Create safe spaces to learn.** Celebrate experimentation, not perfection.
5. **Model the behavior.** Don't just talk about transformation— live it.

When people see you embracing change with openness and calm, they follow. Leadership isn't about being perfect. It's about going first.

What to Say When People Push Back

Here are a few conversation starters to ease tension and open the door to collaboration:

- "Let's test this together and see what we learn."
- "If this gives us back 10 hours a week, how would you use that time?"
- "This doesn't replace your voice. It helps scale it."
- "You don't have to get this right—just get started."

Reconnecting to the Human Side

Ironically, the best way to lead through AI is by doubling down on what makes you human:

- Empathy
- Curiosity

- Honesty
- Vision

People don't need you to have all the answers. They need you to care enough to help them find theirs.

You Are the Emotional Integrator

In a time of transformation, tools matter—but tone matters more.

You're not just the one choosing tech. You're the one choosing how people feel about it. And that determines everything: adoption, momentum, and ultimately results.

The faster you address the emotions under the surface, the faster you move forward with unity and clarity.

You Don't Have to Lead This Transition Alone

The emotional side of AI adoption is often the hardest part—and it's where most leaders get stuck. The good news? You don't have to navigate team resistance, fear, and change management by yourself.

Our Strategic AI Readiness Audit helps you understand not just where to implement AI, but how to lead your team through the emotional journey of transformation.

You'll discover:

- ☑ Your team's specific resistance patterns and how to address them
- ☑ Communication strategies that build excitement instead of fear
- ☑ Which team members to make your AI champions
- ☑ A timeline that honors both business needs and human psychology

Based on your results, you may qualify for a complimentary 20-minute AI Strategy Call where we'll help you craft a leadership approach that gets your entire team moving forward together.

Start your leadership assessment at 10XAIAmplifier.com/audit

Because from here on out, every win with AI isn't just a tech upgrade. It's a trust upgrade.

And the businesses that win in the AI era? They don't just lead with strategy. They lead with safety.

In Chapter 4, we'll move from emotional preparation to practical action—showing you exactly how to start integrating AI in a way that builds confidence and momentum for both you and your team.

Chapter 4
How to Start — Precisely

Your First Five Moves to Integrate AI Without Overwhelm

If you're like most leaders, your question isn't "Should I use AI?" It's "Where the hell do I begin?"

The biggest mistake people make when adopting AI is trying to do everything all at once. They drown in demos, tools, and tutorials—and eventually do nothing.

This chapter is your shortcut around that.

We're not going to throw a hundred options at you. We're going to give you five precise, low-risk, high-impact moves you can make today. No tech team required. No budget blowouts. Just clarity, momentum, and wins you can feel within days.

Let's get you moving.

The Rule: Simplicity Over Sophistication

Before we dive in, here's the principle: **You're not trying to master AI. You're trying to get motion.** Speed beats perfection. And traction builds trust—in the tools and in yourself.

Case Study:
The 5-Step Blueprint Behind Brandon's Transformation

How We Applied the "First Five Moves" Framework to K to Z Interiors

Remember Brandon Barton from Chapter 1? The owner of K to Z Interiors who was drowning in manual processes while hundreds of thousands in quotes sat untouched each month?

We showed you the overview of his transformation—how AI didn't replace his people but amplified them, helping Kelsey reclaim 30% of her day and turning operational chaos into systematic flow.

Now let's pull back the curtain and show you exactly **how** we did it.

What you're about to see is our five-move framework in action— the precise, week-by-week implementation that took Brandon from scattered systems to scalable operations in just 30 days.

This isn't theory. **This is the actual playbook.**

The 30-Day Implementation Timeline

Rather than attempting a complete system overhaul, we implemented our five-move framework progressively, allowing Brandon and his team to build confidence and capability week by week.

Week 1: The Foundation (Move #1 - Personal AI Assistant)

What we implemented:

- Custom GPT trained on K to Z's brand voice, services, and client testimonials
- AI assistant loaded with Google reviews and website content

- Prompt library for client communications and review responses

Brandon's reaction: *"You've done more research on our company than anybody we've ever worked with. The information you pulled out, the links you went to—nobody's ever done that level of preparation."*

Immediate results:
- 50% faster client email responses
- Consistent brand voice across all communications
- Professional Google review responses that reinforced expertise

Week 2: The Documentation Fix (Move #3 - Voice Recording)

The problem we solved: Poor consultation documentation was costing thousands in reorders and client disputes.

Brandon's pain point: *"We had no notes. The customer said they wanted the divider rail at the bottom, but after installation, they claimed they wanted it higher. I'm having to reorder a whole shutter over a miscommunication because we had no record of the original conversation."*

What we implemented:
- Voice recording system for all client consultations
- AI transcription and summary generation
- Integration planning for comprehensive client records

Results:
- Zero reorders due to miscommunication in the following month
- Detailed client preference records for future reference
- Installer confidence in job specifications

Week 3: The Time Reclaim (Move #2 - Task Automation)

The biggest win: We tackled Kelsey's daily administrative burden with a digital end-of-day job reporting system.

What we built:

- Mobile-optimized form accessible via installer iPads
- Photo upload capabilities for before/after documentation (up to 10 photos plus video)
- Conditional fields (clean install vs. problem reporting)
- Automatic routing to client records

The transformation: *"This will save Kelsey 40 minutes every day,"* Brandon calculated. *"That's over 3 hours saved per week, just from one simple automation."*

Results:

- **40+ minutes daily** saved on administrative tasks
- Complete job documentation with photo evidence
- Clear communication channel between field and office
- Foundation for installer performance tracking and incentives

Week 4: The Visibility Boost (Move #4 - Decision Dashboard)

What we implemented:

- Automated lead tracking and follow-up scheduling
- Google My Business optimization and review management system
- Email and SMS cadence automation
- Performance reporting dashboard

Results:

- Clear visibility into lead conversion pipeline
- Systematic review generation reaching toward the 100-review milestone

- Automated prospect nurturing reducing manual follow-up burden
- Data-driven insights replacing gut-feel decision making

Six Months Later: The Compound Effect

Operational Transformation:

- **Kelsey reclaimed 30% of her workday** (2.5+ hours daily)
- **Clean install documentation became standard practice**
- **Client communication became consistent and professional**
- **Follow-up processes became automatic, not accidental**
- **Review generation hit systematic rhythm** (targeting 2 reviews weekly toward 100-review SEO milestone)

Business Impact:

- **Improved conversion rates** on quoted work through better follow-up
- **Reduced reorders and disputes** through comprehensive documentation
- **Enhanced client satisfaction** leading to more referrals and positive reviews
- **Scalable foundation** supporting growth without proportional administrative burden increase

Brandon's Bottom Line: *"I wish I'd met you three months ago. I would have never gone with the digital company I have now. You've shown me exactly what I needed to do—questions I've been asking everybody I've dealt with for years, and nobody's been able to provide this level of answers."*

The Key Success Factors

What made K to Z's transformation work when so many AI implementations fail?

1. **Started with pain, not possibility:** We identified Kelsey's daily 40-minute administrative burden rather than chasing shiny AI features
2. **Built progressively:** Each week's implementation built on the previous week's foundation, preventing overwhelm
3. **Maintained human oversight:** AI enhanced human decision-making rather than replacing it entirely
4. **Focused on operational relief:** Every implementation directly reduced friction in existing workflows
5. **Measured real impact:** Time saved, reorders prevented, and client satisfaction improvements were tracked and celebrated

The Five Moves: Your Implementation Guide

Now that you've seen exactly how it works in practice, here are the five moves you can implement starting today:

Move #1: Train a Personal Prompt Assistant

Create your go-to AI for writing, planning, and idea generation.

Use ChatGPT, Claude, or another LLM to build a "prompt profile" that understands your tone, your goals, and your audience. Think of it as a junior strategist who gets smarter every time you work together.

- **Detailed Implementation:**

1. **Gather your brand materials:** Website copy, recent emails, client testimonials, service descriptions
2. **Create your training prompt:**

☐

"Act as my business development strategist for [YOUR COMPANY].

Here's our brand voice [PASTE SAMPLE CONTENT].

Our services include [LIST SERVICES].

Our ideal clients are [DESCRIBE AUDIENCE].

Help me generate email copy, review responses, and client communication that matches this tone and expertise level."

☐

3. **Test with real scenarios:** Use it to draft a client follow-up email, respond to a review, or create social media content
4. **Refine based on results:** Adjust the prompt based on what works and doesn't work

⏱ **Time to implement:** 15-30 minutes

✳ **ROI:** Immediate copy support, better brainstorming, no blank pages

Success metric: You're drafting client communications 50% faster with more consistent messaging

Move #2: Automate One Recurring Task with AI + Zapier

Eliminate one mindless task you hate doing every week.

Pick something repetitive—like sorting leads, generating summaries, or filing documents—and automate it with a tool like Zapier, Make, or Pabbly.

● **Detailed Implementation:**

1. **Identify your biggest time drain:** What recurring task takes 30+ minutes weekly and follows a predictable pattern?
2. **Map the current process:** Document each step (Email arrives → Read → Categorize → File → Follow up)
3. **Build the automation:**
   o Connect your email to AI analysis tool

- o Set up automatic categorization and filing
- o Create trigger for follow-up actions
4. Test with small batch: Run 10-20 items through the system before full deployment

Real Example from K to Z: "New installer report submitted → AI categorizes as clean install or problem → Photos auto-filed to client record → Kelsey gets summary notification instead of managing individual text threads"

🕐 **Time to implement:** 30–60 minutes

❋ **ROI:** Saves hours weekly, frees mental bandwidth

Success metric: You're spending 5+ fewer hours weekly on administrative tasks

Move #3: Record, Transcribe, and Repurpose with AI

Turn your spoken thoughts into scalable content.

Use tools like Otter.ai, Descript, or Castmagic to record your voice, transcribe it, and turn it into multiple assets (blogs, emails, social posts). You already know the content—you just need to unlock it from your head.

⚪ **Detailed Implementation:**

1. **Choose your recording method:** Phone app, dedicated device (like Plaud), or computer software
2. **Start with client calls:** Record consultation calls (with permission) for better documentation
3. **Set up transcription workflow:**
   - o Upload to transcription service
   - o Use AI to create summary and action items
   - o Generate follow-up communications and client records
4. **Expand to content creation:** Record 10-minute voice memos and turn them into multiple content pieces

Real Example from K to Z: Brandon wanted to use Plaud AI to record client consultations, then automatically transcribe and summarize them into client records to prevent costly reorders due to miscommunication.

⏱ **Time to implement:** 1 hour setup

✴ **ROI:** Multiplies your output with zero extra writing, better client documentation

Success metric: You're creating 3x more content in the same time, with better client records

Move #4: Create a "Decision Dashboard"

Use AI to summarize your data so you can lead, not sift.

AI can surface trends, flag outliers, and recommend next steps. You don't need a full data science team—you just need better visibility.

◉ **Detailed Implementation:**

1. **Gather your key data sources:** Sales reports, customer feedback, website analytics, financial data
2. **Create analysis prompts:**

☐

```
"Analyze this sales data and tell me:

- What are the three biggest trends?

- What surprises you?

- What would you optimize first?

- What risks do you see?"
```

☐

3. **Set up regular review rhythm:** Weekly or monthly data analysis sessions
4. **Create action item tracking:** Use AI insights to drive specific business decisions

Real Example from K to Z: We created a dashboard showing lead conversion rates, review generation progress (tracking toward 100-review SEO milestone), and installer performance metrics, giving Brandon clear visibility instead of gut-feel decision making.

⏱ **Time to implement:** 30 minutes

✳ **ROI:** Smarter decisions, faster pivots, clearer priorities

Success metric: You're making data-driven decisions 80% faster with higher confidence

Move #5: Design a 90-Day Learning Loop

Make AI a rhythm, not a resolution.

Block 1 hour a week to explore, test, and document what works. AI isn't a one-time download—it's a moving stream. Your job is to stay in flow without getting swept away.

• **Detailed Implementation:**

1. **Schedule weekly "AI Lab" time:** Same day/time each week, non-negotiable
2. **Create experimentation framework:**
   o Week 1: Try one new tool or prompt
   o Week 2: Document what worked/didn't work
   o Week 3: Scale successful experiments
   o Week 4: Share learnings with team
3. **Build knowledge repository:** Keep notes on successful prompts, useful tools, and lessons learned
4. **Measure compound results:** Track time saved, revenue generated, or processes improved

Real Example from K to Z: Brandon committed to weekly AI exploration sessions, building on the foundation we established and continuously finding new applications for his specific business challenges.

⏱ **Time to implement:** 5 minutes to schedule

✳ **ROI:** Compounds learning, prevents stagnation, builds internal culture of innovation

Success metric: You're consistently finding new AI applications and your team is actively contributing ideas

How to Know You're Winning

You're not winning with AI when everything's automated.

You're winning when:

- Your calendar has margin
- Your team has clarity
- Your content isn't backed up in your head
- Decisions move faster and stress drops lower
- You're spending time on strategy, not just execution

That's the early win stack. That's traction. And it builds from here.

Momentum Over Mastery

You don't need to use every tool. You don't need to understand every algorithm. You need to **start**. You need to **stack wins**. You need to **build trust** in your own ability to adapt.

And these five moves? They're the beginning of that journey.

In the next chapter, we'll show you what to use AI for right now—with plug-and-play prompts, workflow ideas, and some of the most powerful use cases we've seen create real-time wins for businesses just like Brandon's.

But for now—pick one move above and do it **this week**.

That one action plants your flag in the future. You're not watching the AI wave pass by anymore.

You're riding it.

Chapter 5
What Should I Use AI For —
Right Now?

The Fastest Wins You Can Plug In This Week

Let's kill the myth that AI is only for long-term transformation.

Yes, AI can rebuild your systems. Yes, it can scale your operations. But what most business owners really need is a few immediate wins. They need to feel what it's like when something finally clicks. When a task gets easier. When their time suddenly reappears.

This chapter is where that happens.

We're going to break down high-impact, low-friction ways to integrate AI this week. No fluff. No complexity. Just fast ROI across five key areas:

1. Writing & Content Creation

Stop staring at blank pages. Start with smarter prompts.

Whether it's email, social, blogs, or sales pages—AI can get you to a draft in minutes.

📍 **Try this prompt:** "Act as my marketing assistant. Write a social post to help [ideal client] solve [common pain point], using a voice that's [casual/professional/witty], and include a soft CTA to book a call."

🔥 **PRO TIP:** Use the "10 Variations" approach. Generate one solid piece of content, then ask: "Give me 10 different angles on this same topic." Batch schedule them using Buffer or Later.

Real Example:

```
☐"Act as my marketing assistant for K to Z Interiors.


Our brand voice is: Professional but approachable,
educational without being condescending.


Our ideal clients are: Homeowners in Louisiana
planning interior updates who value quality and
expertise.


Write a LinkedIn post to help them understand the
difference between blinds, shades, and shutters,
using a helpful tone, and include a soft CTA to book
a complimentary consultation."☐
```

💥 **Outcome:** More visibility. Less mental fatigue. 5x content creation speed.

2. Email Management & Communication

Clear your inbox without clearing your calendar.

AI can summarize threads, draft replies, categorize responses, and even automate follow-ups.

💡 **Try this tool combo:**

- Gmail + ChatGPT via Zapier to auto-summarize new leads
- Superhuman or Missive for smart replies that save your tone

Power Prompt for Email Summaries:

☐"Summarize this email thread and tell me:

1. What action is needed from me?

2. What's the deadline?

3. Who else is involved?

4. Draft a professional response that acknowledges their concern and moves things forward"☐

💧 **PRO TIP:** Set up automated email categorization. New email arrives → AI scans content → Auto-tags as "Urgent," "Follow-up," or "Information Only" → Filtered into folders automatically.

Real Result from K to Z: Lead response time dropped from hours to minutes, conversion rates increased by 23%.

✳ **Outcome:** Regain hours. Stay responsive. Look brilliant.

3. Market Research & Audience Intel

Know exactly what your customers want—without hours of digging.

AI can analyze reviews, survey results, or support tickets to surface what your audience is really asking for.

◉ **Try this process:** Drop your customer survey or testimonials into ChatGPT and ask:

- "What are the top 5 problems people mention?"
- "What language do they use to describe their pain?"
- "What objections show up most?"

Advanced Analysis Prompt:

☐"Analyze these customer reviews and tell me:

1. What are the top 5 problems people mention?

2. What language do they use to describe their pain?

3. What objections show up most frequently?

4. What do they love most about our service?

5. What gaps do you see in our current offerings?" ☐

⬤ **PRO TIP:** Use the exact language customers use in your marketing copy. When prospects see their own words reflected back, they think "It's like you read my mind."

Case Study Success: A consulting firm analyzed 200+ testimonials, discovered clients valued "peace of mind" over "expertise," rewrote their website copy, and saw a 40% increase in consultation bookings.

✴ **Outcome:** Better offers. Clearer messaging. Higher conversions.

4. Customer Experience & Personalization

Make your audience feel seen—without adding headcount.

You can use AI to tailor onboarding, segment messaging, and even write personalized responses at scale.

🍢 **Try this flow:** Lead submits a form → AI writes a tailored welcome email based on their response → Email is auto-sent with a dynamic video or CTA

Personalization Template:

☐"Create a personalized welcome email for a [CLIENT TYPE] who just [TOOK SPECIFIC ACTION].

Include:

- Acknowledgment of their specific situation

- 2-3 relevant resources

- Clear next steps

- Warm, professional tone that builds trust"☐

💧 **PRO TIP:** Create behavioral triggers. Website visitor downloads specific resource → Triggered email sequence about that topic. Client completes project → Automated review request with personalized message.

Real Implementation: Marketing agency reduced onboarding time by 50% and improved client satisfaction scores by 85% using AI-powered personalized onboarding sequences.

✳ **Outcome:** More intimacy. Less time. Higher retention.

5. Strategy & Decision Support

You don't need more data. You need smarter decisions.

AI can read your data, find patterns, and recommend next steps—so you lead with vision, not guesswork.

💧 **Try this prompt:** "Review this sales call transcript. What were the top objections? Where did the prospect seem most engaged? What would you improve?"

💧 **Or upload last quarter's financials and ask:** "What stands out? Where can we cut cost or increase margin?"

Advanced Decision Support Prompt:

☐"Analyze this financial data and identify:

1. What trends stand out over the past 6 months?

2. Where are we spending inefficiently?

3. What revenue opportunities are we missing?

4. What should be our top 3 priorities next quarter?

5. What risks should we prepare for?" ☐

💧 **PRO TIP:** Create a weekly business review process. Monday: Upload key metrics. Wednesday: Review AI insights with team. Friday: Implement top 2-3 recommended actions.

✳ **Outcome:** Faster pivots. Fewer blind spots. Confident leadership decisions.

Another "Fast Five" for AI Wins

These are your quick-hit prompts and workflows you can screenshot or swipe:

1. **"Write a 500-word blog post based on this voice memo."** → Turns spoken thought into publishable content.
2. **"Draft three subject lines and two versions of a follow-up email for this offer."** → Saves mental energy, improves open rates.
3. **"Summarize this customer feedback into three product improvement ideas."** → Informs roadmap without guesswork.
4. **"Give me 10 engaging questions to ask my audience on Instagram about [topic]."** → Drives engagement. Builds trust.
5. **"Build a 30-day onboarding sequence for this type of customer journey."** → Automates nurture while feeling deeply personal.

Implementation Strategy: The 3-2-1 Approach

Don't try to implement everything at once. Use this progressive approach:

Week 1: Pick **3** prompts from the Fast Five and test them
Week 2: Choose **2** from the five main categories that solve your biggest pain points
Week 3: Implement **1** complete workflow that connects multiple AI tools

This builds competence, confidence, and compound results.

A Word About Overthinking

Don't fall into the trap of trying to "do it right." Do it real. Do it now.

Every action builds capability. Every use builds fluency. Every result builds belief.

Your goal this week: pick three of the use cases above and implement them. Not perfectly. Just decisively.

Because you don't get momentum by studying AI.

You get momentum by moving with it.

In the next chapter, we'll zoom out to show you how to structure your business model for amplification—how to design systems and strategy around AI so you're not just saving time, you're creating scalable flow.

But first, go create a few quick wins. Then come back ready to scale.

Chapter 6
Your Amplified Business Model

Designing a Business That Scales Without You as the Bottleneck

The most dangerous myth in business? That more effort equals more growth.

At some point, adding hours stops adding revenue. Hiring more people just adds friction. Your time becomes a liability. Your mind becomes a bottleneck.

Enter the Amplified Business Model—where AI replaces friction with flow, and your systems scale without burning you out.

This chapter is about rebuilding the foundation of your business around leverage, not labor. You'll discover how to shift from a hustle-based model to one where intelligence—not intensity—does the heavy lifting.

The Shift: From Effort-Based to Intelligence-Based Growth

In traditional businesses, growth looks like this:
- Add more clients → add more hours
- Launch more products → increase complexity

- Hit a ceiling → hire more staff

In amplified businesses, growth looks like this:
- Add more value → automate delivery
- Launch strategically → optimize based on data
- Hit a ceiling → redesign the system, not the person

The difference isn't in what you sell. It's in how you scale.

Real-World Transformation:
The Coaching Business Revolution

Before we dive into the framework, let's see how one coach completely redesigned her business model using AI amplification.

Meet Sarah: From Overwhelmed to Exponential

Before the Amplified Model: Sarah was a successful leadership coach earning six figures, but she was trapped. Working 60+ hours a week, she felt like she was running a job, not a business.

Her daily reality:
- Manually handled all intake forms and onboarding emails
- Wrote every piece of content from scratch, managed to blog once a month
- Personally summarized every client call and sent follow-up emails
- Felt "at capacity" serving just 12 high-end clients
- Couldn't take a real vacation without her business suffering

After Implementing the Amplified Model: Six months later, Sarah was serving 30 clients, working 40 hours a week, and had taken three week-long vacations.

Her transformation:
- **AI intake system** auto-onboarded new clients with personalized welcome sequences

- **Call recordings** were automatically summarized into action steps and emailed via automation
- **AI content engine** turned her weekly 20-minute video into 6+ social posts, 2 emails, and 1 blog post
- **Smart scheduling system** optimized her calendar and handled rescheduling
- **Automated client check-ins** maintained connection between sessions

The Results: 150% more clients, 33% fewer working hours, 200% more content output.

That's not just efficiency. That's expansion.

Start With the Core: Where Are You the Bottleneck?

The first step to amplification is identifying where you're still central to survival.

Ask yourself:
- What tasks collapse if I step away for 2 weeks?
- What problems only I can solve today?
- What systems still rely on memory or manual effort?

The Bottleneck Audit Exercise:

Track your time for one week and categorize every task as:
- **High-Value, High-Touch** (only you can do it)
- **High-Value, Low-Touch** (important but systemizable)
- **Low-Value, High-Touch** (you're doing it but shouldn't be)
- **Low-Value, Low-Touch** (eliminate or automate immediately)

 🔥 **PRO TIP:** Use AI to help with this audit. Record voice memos throughout your day describing what you're doing, then ask AI to categorize and analyze patterns.

Highlight your friction zones. They're the first to be redesigned with AI.

The Three Leverage Layers

To build an AI-amplified model, think in layers—not just tools.

1. Time Leverage

Replace repeatable actions with automations and assistants.

- **Examples:**
- Lead follow-up sequences that adapt based on engagement
- Meeting summaries generated from recordings
- Inbox triage that prioritizes and categorizes automatically
- Social media content that publishes on optimal schedules

Real Implementation: A real estate agent used AI to automate her entire lead nurture process. New leads received personalized market reports, property alerts, and educational content automatically. Result: 40% more deals closed with 50% less manual effort.

2. Knowledge Leverage

Extract what's in your brain and turn it into systems, training, and assets.

- **Examples:**
- Client onboarding templates that adapt to different business types
- FAQ systems that learn from every customer interaction
- Sales scripts that evolve based on successful conversations
- Standard Operating Procedures created from your voice explanations

Knowledge Extraction Process:

Step 1: Record yourself explaining a process (10-15 minutes)

Step 2: AI transcribes and structures the content

Step 3: AI creates step-by-step guides, checklists, and training materials

Step 4: System learns and improves from each use☐

3. Energy Leverage

Spend more time in your zone of genius by offloading what drains you.

Examples:

- AI writing tools handle first drafts, you focus on strategy
- Scheduling bots eliminate back-and-forth calendar coordination
- Content batching systems multiply your creative output
- Automated reporting gives you insights without data analysis

This isn't about doing less work. It's about doing the right work—and multiplying its reach.

Build Your Amplification Stack

Here's a detailed blueprint to restructure your business model using AI-enhanced components:

Business Function	Traditional Approach	AI-Enhanced Solution	Amplified Outcome
Lead Gen & Outreach	Manual prospecting, cold calling	AI email sequencers + audience analyzers + behavioral triggers	More qualified leads, faster conversion, personalized at scale
Sales & Follow-up	Manual call notes, spreadsheet tracking	AI call summarizers + smart CRMs + predictive scoring	Better conversion, no lost deals, data-driven priorities

Business Function	Traditional Approach	AI-Enhanced Solution	Amplified Outcome
Content & Communication	Write everything from scratch	Prompt libraries + repurposing tools + voice-to-content systems	High-output visibility with low input, consistent brand voice
Client Delivery	Manual check-ins, reactive support	AI dashboards + automated touchpoints + predictive insights	More value delivered, proactive service, reduced churn
Operations	Manual processes, memory-based systems	SOP automation + workflow optimization + performance tracking	Predictability, scalability, peace of mind

The Amplified Business Blueprint: Step-by-Step Implementation

Phase 1: Foundation (Weeks 1-2)

Goal: Establish basic AI assistance and data collection

Actions:

- Set up AI writing assistant with your brand voice
- Implement basic email automation for new leads
- Create voice recording system for knowledge capture
- Begin tracking all business processes

Success Metric: 5+ hours weekly time savings

Phase 2: Systematization (Weeks 3-6)

Goal: Convert manual processes into AI-enhanced workflows

Actions:

- Build client onboarding automation
- Create content repurposing engine

- Implement meeting summary and follow-up system
- Design basic performance dashboards

Success Metric: 50% reduction in manual administrative tasks

Phase 3: Optimization (Weeks 7-12)

Goal: Refine systems and scale operations

Actions:

- Advanced personalization and segmentation
- Predictive analytics for business decisions
- Team training on AI-enhanced workflows
- Continuous improvement loops

Success Metric: 2x capacity with same or fewer working hours

Don't Just Delegate Tasks. Delegate Thinking.

AI isn't just about doing your to-dos faster. It's about upgrading how your business thinks.

Traditional Thinking: React to problems as they arise
Amplified Thinking: Predict and prevent problems before they happen

Examples of Delegated Thinking:

- Let AI surface patterns in customer behavior you might miss
- Have it make recommendations based on performance data
- Allow it to optimize touchpoints based on engagement analytics
- Use it to identify opportunities and risks in your market

Case Study: The Predictive Service Business

A consulting firm implemented AI thinking across their operations:

- **Client Success Prediction:** AI analyzed communication patterns and project metrics to identify clients at risk of churn
- **Pricing Optimization:** AI recommended pricing adjustments based on market data and client value patterns
- **Capacity Planning:** AI predicted busy periods and recommended staffing adjustments
- **Content Strategy:** AI identified which topics resonated most with their audience

Results: 30% increase in client retention, 25% improvement in profit margins, 50% more strategic decision-making accuracy.

You're not outsourcing leadership. You're extending it.

Another Fast Five AI Wins

Ready-to-implement strategies for business model amplification:

1. **"Create a client onboarding automation that personalizes based on their industry and goals."** → Scalable intimacy from day one.
2. **"Build a content multiplication system that turns one piece into 10+ assets."** → Maximum visibility with minimal effort.
3. **"Design a predictive client success system that flags potential issues early."** → Proactive service that prevents churn.
4. **"Implement an AI-powered performance dashboard that provides weekly business insights."** → Data-driven decisions without analysis paralysis.

5. **"Create an automated referral system that nurtures past clients into advocates."** → Systematic growth through relationship leverage.

Your Next Step: The Business Audit Sprint

Complete this rapid assessment to identify your amplification opportunities:

Step 1: List every activity you do weekly
Step 2: Highlight what:
- Repeats (same process, different inputs)
- Drains (tasks you avoid or procrastinate)
- Delays others (bottlenecks you create)

Step 3: Choose three to offload or optimize with AI in the next 30 days

🔥 **PRO TIP:** If it's not high-trust or high-creativity, it can probably be systemized.

The Amplification Priority Matrix:

High Impact, Low Effort	High Impact, High Effort
Start here → Quick wins that build momentum	Plan these → Major transformations
Low Impact, Low Effort	**Low Impact, High Effort**
Automate these → Free up mental space	Eliminate these → Pure waste

Case Study: The Agency That Doubled Down

The Challenge: A marketing agency was stuck at $500K revenue with 8 team members working 50+ hour weeks.

The Amplified Approach:
- **Client Reporting:** AI analyzed campaign data and generated branded reports automatically

- **Content Creation:** AI helped create first drafts for all client content, team focused on strategy and refinement
- **Project Management:** AI tracked project progress and automatically sent updates to clients
- **Business Development:** AI identified and qualified potential clients from multiple data sources

The Results After 6 Months:

- Revenue grew to $750K with the same team size
- Average work week dropped to 40 hours
- Client satisfaction scores increased 35%
- Profit margins improved by 20%
- Team stress levels significantly decreased

The Key Insight: They didn't replace their expertise—they amplified it.

Freedom Is a Design Decision

Most entrepreneurs build freedom-hungry businesses that actually trap them.

Amplified businesses are different. They're designed to:

- Run without you in every loop
- Scale without compromising quality
- Operate with fewer people, more precision, and way less drama

The Freedom Test: Can your business operate for 2 weeks without you? If not, you haven't built a business—you've built a job with more responsibility.

The Amplification Promise: AI doesn't just make your business more efficient. It makes it more intelligent, more responsive, and more valuable—while giving you back your life.

And AI is the engine that makes this real—today, not next year.

In the next chapter, we'll go deep on your AI tech stack—how to choose the right tools, avoid the noise, and make everything work together without turning into IT support.

But for now, remember: Your bottleneck isn't a problem. It's a blueprint.

Let's redesign it—intelligently.

Chapter 7
Your AI Tech Stack Made Simple

Choose the Right Tools Without Turning into a Tech Department

You don't need 37 AI tools.

You need 3 to 5 that actually talk to each other—and help you talk to your audience, team, and prospects more clearly and effectively.

The biggest trap business owners fall into right now? **Tool chasing.**

They collect software like it's Pokémon cards, thinking each new one is the missing piece.

Here's the truth: a great stack isn't about volume. It's about **intentionality.**

This chapter will show you exactly how to build one that works with your brain and for your business.

What Is an AI Stack, Really?

Your AI tech stack is the collection of tools, automations, and workflows that allow you to:

- Save time
- Make decisions
- Personalize communication
- Scale delivery
- Maintain consistency without micromanaging

The best stacks act like **amplifiers**: they take your thinking and scale it through systems. The worst ones act like **friction**: more clicks, more confusion, more regret.

The Tool Trap: Why More Isn't Better

We've seen the same pattern repeat with countless business owners. They start with one AI tool, get excited by the possibilities, then begin collecting every new platform that promises to solve their problems.

The Typical Downward Spiral:

- 10+ different AI tools (each promising something unique)
- Multiple automation platforms (forgotten subscriptions)
- Dozens of content creation apps (for every possible format)
- Various analytics dashboards (each showing different metrics)
- Browser bookmarks full of "promising" solutions

The Predictable Results:

- Hours daily just managing tools
- Data scattered across platforms with no single source of truth
- Monthly software costs spiraling out of control
- Team confusion about which tool to use when
- Analysis paralysis instead of action

The Solution: A strategic framework that prioritizes function over features.

The 5-Category Framework That Works

Every successful AI stack we've analyzed falls into five core categories. Master these categories, and you'll avoid the tool trap while building something that actually amplifies your business.

1. Thinking & Writing

Your AI brain—idea generation, outlines, messaging, and strategy.

What it does: Handles content creation, brainstorming, strategic planning, and communication drafting.

Why you need it: Writing is thinking made visible. An AI writing assistant doesn't just save time—it helps you think more clearly and communicate more effectively.

Selection criteria:
- Can it learn your brand voice?
- Does it integrate with your other tools?
- Is the interface intuitive for daily use?
- Can your team access and use it consistently?

2. Automation & Workflow

Your AI hands—integrating apps, automating tasks, routing data.

What it does: Connects your business tools, automates repetitive processes, and ensures data flows smoothly between systems.

Why you need it: Every manual handoff is a potential bottleneck. Automation eliminates the friction between your tools and your outcomes.

Selection criteria:

- How many of your current tools does it connect?
- Can you build workflows without coding?
- Is the pricing structure sustainable as you scale?
- Does it have reliable customer support?

3. Media & Content Creation

Your AI mouth—video, voice, visuals.

What it does: Transforms your ideas into multiple content formats, repurposes existing content, and creates visual assets.

Why you need it: Content multiplication is the key to consistent visibility without constant creation.

Selection criteria:

- Does it maintain your brand consistency?
- Can it handle your primary content formats?
- Is the learning curve manageable for your team?
- Does the output quality meet your standards?

4. Data & Decision Support

Your AI eyes—analyzing numbers, surfacing insights, flagging patterns.

What it does: Analyzes your business data, identifies trends, surfaces actionable insights, and supports strategic decision-making.

Why you need it: Data without insights is just noise. AI turns your business metrics into strategic intelligence.

Selection criteria:

- Can it access your key data sources?
- Does it provide actionable recommendations, not just reports?
- Is the interface designed for business leaders, not data scientists?
- Can it adapt to your specific business model and metrics?

5. Customer Experience & Personalization

Your AI handshake—engaging, onboarding, serving.

What it does: Personalizes customer interactions, automates support responses, and creates dynamic experiences based on customer behavior.

Why you need it: Personal attention doesn't have to require personal time. AI can deliver customized experiences at scale.

Selection criteria:

- Does it integrate with your CRM and customer data?
- Can it maintain your brand voice in customer interactions?
- Is it sophisticated enough to handle your customer complexity?
- Does it escalate appropriately to human team members?

The Stack Builder Rules

Rule #1: Fit First, Flash Later

Don't start with what's new. Start with what fits:

- What platforms do you already use?
- What needs to talk to each other?
- What task, if automated, would save you 10+ hours a month?

That's your starting point. Build outward from there.

Rule #2: One Tool Per Category (Until Proven Otherwise)

Resist the urge to have backup tools for every function. Master one tool completely before considering alternatives. Tool redundancy creates complexity, not capability.

Rule #3: Integration Over Isolation

Every tool you choose should connect to at least one other tool in your stack. Isolated tools create data silos and workflow friction.

Rule #4: Document Everything

Create simple documentation for each tool:

- What it does in plain English
- Who on your team uses it
- How it connects to other tools
- What triggers its use

The Integration Assessment

Before adding any new tool to your stack, run it through this filter:

Technical Fit:
1. Does it integrate with my existing systems?
2. Will it reduce my overall tool count?
3. Can my team learn it in under 2 hours?

Business Fit: 4. Does it solve a real problem or just look cool? 5. Is the ROI measurable within 30 days? 6. Will it scale with my business growth?

Strategic Fit: 7. Does it align with my business priorities? 8. Will it create more connection or more complexity? 9. Can I afford to maintain it long-term?

Only add tools that score well across all three categories.

Avoiding Stack Overwhelm

You're not building a tech empire. You're building a machine that gives you back your mind.

The Lean Stack Principles:

Visual Mapping: Create a simple diagram showing how your tools connect. If you can't draw it clearly, it's too complex.

The 5-Tool Rule: Most businesses need no more than 5 core tools. If you have more than 8, you probably have redundancy or tools you're not actually using.

The Handoff Test: Every tool should either receive input from another tool or send output to another tool. Isolated tools are usually unnecessary tools.

The Team Test: If a new team member can't understand your stack in 30 minutes, simplify it.

The Stack Evolution Framework

Your AI stack should evolve with your business, not your curiosity:

Stage 1: Foundation (Months 1-6)

Focus: Basic automation and AI assistance
Goal: Prove ROI and build confidence
Tools: 3-5 core tools, simple integrations
Success Metric: 10+ hours weekly time savings

Stage 2: Integration (Months 6-12)

Focus: Connecting systems and improving workflows
Goal: Eliminate manual handoffs and data silos
Tools: Advanced automation, data connections

Success Metric: 90% of routine tasks automated

Stage 3: Intelligence (Months 12+)

Focus: Predictive insights and advanced personalization
Goal: Proactive business optimization and competitive advantage
Tools: AI-powered analytics, machine learning features
Success Metric: Strategic decisions made 3x faster

The Monthly Stack Audit

Every month, review your tools with this brutal clarity test:

Value Questions:

- Is this saving me time?
- Is this making me smarter?
- Is this creating more connection?

Usage Questions:

- Did I actually use this tool this month?
- What measurable impact did it have?
- Would my business suffer if it disappeared tomorrow?

Integration Questions:

- Is this tool talking to my other tools?
- Is it creating data silos or solving them?
- Does my team know how and when to use it?

If any tool fails these tests, eliminate it. Complexity is a tax on creativity.

Another Fast Five AI Wins

Strategic stack decisions you can make immediately:

1. **"Map your current tools visually and eliminate any that don't connect to others."** → Instant simplification and cost savings.
2. **"Choose one category from the 5-part framework and optimize it completely."** → Deep competence beats shallow coverage.
3. **"Document your tool stack in plain English for new team members."** → Clarity that scales with your business.
4. **"Set a monthly calendar reminder for your stack audit."** → Systematic optimization instead of reactive changes.
5. **"Apply the integration assessment to any tool you're considering."** → Intentional decisions instead of impulse additions.

Build for Momentum, Not Mastery

Don't try to build a perfect stack out of the gate.

Build a smart **first version**. Use it. Refine it. Add only what supports your amplified goals.

The 30-Day Stack Challenge:

- **Week 1:** Audit your current tools using the 5-category framework
- **Week 2:** Eliminate redundant or unused tools
- **Week 3:** Create one integration between your two most-used tools
- **Week 4:** Document your streamlined stack and plan your next evolution

Remember: **Tech doesn't make you powerful. Clarity does.**

Your Stack Blueprint Awaits

The strategic framework you just learned is timeless. But the specific tools, pricing, and setup instructions change rapidly in the AI space.

Rather than lock you into recommendations that might be outdated by the time you read this, we've created a living resource that stays current with the best available tools.

Download the complete AI Tech Stack Builder with current tool recommendations, pricing comparisons, setup guides, and integration tutorials at **10XAIAmplifier.com/stack**

You'll get:

- ☑ Current tool recommendations for each category
- ☑ Stack templates by business size and budget
- ☑ Step-by-step setup and integration guides
- ☑ Monthly updates as the landscape evolves
- ☑ Access to our AI Officer community for support

In the next chapter, we'll show you how to get your team on board without resistance—because an AI-powered business still needs human trust and buy-in to scale.

But for now, apply the framework. Simplify your stack. And remember:

The best stack is the one you actually use consistently.

Chapter 8
Your Team, Your Culture, and the Buy-In Blueprint

You can't automate trust.

You have to earn it—and protect it—especially in the middle of transformation.

AI might be reshaping industries at warp speed, but in your business? The real transformation starts with your team. Because even the best tech stack will stall if your people don't believe in the direction you're heading.

This chapter is your blueprint for building a culture that adopts AI with clarity and confidence—not confusion or fear.

The Real Reasons Teams Resist AI

Let's not sugarcoat it: AI makes people nervous. But that anxiety isn't random—it's rooted in real, emotional concerns that smart leaders anticipate and address.

Here are the 10 most common resistance points your team might be feeling:

Emotional Trigger	Underlying Fear
1. Fear of Replacement	"Will I lose my job?"
2. Loss of Control	"Am I still relevant in this process?"
3. Overwhelm from Complexity	"This sounds too technical for me."
4. Fear of Looking Incompetent	"I don't want to mess up or look dumb."
5. Lack of Context	"Why are we doing this again?"
6. Mistrust of Motives	"Is this just about cutting costs?"
7. Emotional Attachment to Old Systems	"But this is how we've always done it."
8. Fear of Losing Human Touch	"Is this going to make us robotic?"
9. Inequity in Access or Support	"Why do they get training and we don't?"
10. Change Fatigue	"Another new thing? I'm exhausted."

How to Reframe AI for Team Buy-In

You don't shift resistance by force. You shift it with story—clear, compassionate narratives that reposition AI as a source of empowerment, not threat.

Resistance	Reframe Strategy	Leadership Language
1. Fear of Replacement	Position AI as a support system	"It's here to help you do more of what only you can do."
2. Loss of Control	Co-create solutions	"You get to shape how this evolves."
3. Complexity Overwhelm	Focus on one tool at a time	"One win at a time. We're not changing everything overnight."

Resistance	Reframe Strategy	Leadership Language
4. Fear of Incompetence	Normalize imperfection	"None of us are experts—we're all learners here."
5. Lack of Context	Clarify the "why"	"This helps us serve better and grow smarter."
6. Mistrust of Motives	Be radically transparent	"Here's what we're doing—and what we're not doing."
7. Attachment to Old Ways	Honor the past	"What worked before brought us here. Now we evolve forward."
8. Human Touch Fear	Reposition AI as connection amplifier	"AI helps us be more human at scale."
9. Perceived Inequity	Democratize access	"Everyone gets a chance to learn, grow, and explore."
10. Change Fatigue	Lead with relief	"We're not adding work—we're removing friction."

The 5-Stage Buy-In Blueprint

1. **Acknowledge the Emotion** Start by letting your team speak. Listen more than you talk.
2. **Clarify the Why** Tie AI back to what they value: purpose, impact, time.
3. **Co-Create Change** Don't impose systems. Build them with the people who'll use them.
4. **Deliver a Quick Win** Show them what gets easier—not just what changes.
5. **Empower Ongoing Curiosity** Appoint AI champions. Offer micro-trainings. Keep the learning loop open.

Team Activation Exercise: Meet Your New Thinking Partner

Want to move from fear to fluency? Let your team experience AI—not just hear about it.

Host a 45-minute session that lets everyone get hands-on with ChatGPT or Claude. Let them pick real prompts, generate outputs, and debrief together.

🎯 Goals:

- Build confidence
- Spark curiosity
- Uncover high-impact use cases

💼 What You'll Need:

- Internet-connected devices
- AI account access (free is fine)
- Our Team AI Starter Pack (includes role-specific prompts, worksheets, and onboarding scripts)

📅 Need Help With Team Training?

Ready to run a team AI workshop but want expert guidance? Our Strategic AI Readiness Diagnostic includes team readiness assessment and training resources. Based on your results, you may qualify for a complimentary 20-minute AI Strategy Call where we can help you design and facilitate team training sessions that build excitement instead of resistance.

Get your team readiness diagnostic at
10XAIAmplifier.com/diagnostic

Culture Eats Automation for Breakfast

When your team believes the story, they build the system. When they own the tools, they scale the outcomes. And when they feel safe, seen, and supported—they move faster than any AI ever could.

Master Team Leadership in the AI Era

Leading a team through AI transformation requires more than just technical knowledge—it requires emotional intelligence, change management skills, and the ability to build trust while navigating uncertainty.

You don't have to figure out team dynamics, resistance patterns, and cultural change by yourself.

Our Strategic AI Readiness Diagnostic includes a comprehensive team leadership assessment that helps you:

- ☑ Identify which team members will be your early adopters vs. skeptics
- ☑ Understand the specific concerns and motivations of each team member
- ☑ Develop customized communication strategies for different personality types
- ☑ Create a timeline that honors both business goals and human psychology
- ☑ Build internal champions who can help drive adoption

Based on your results, you may qualify for a complimentary 20-minute AI Strategy Call where we'll help you craft a team leadership approach that gets everyone moving forward together with confidence and clarity.

Start your team leadership diagnostic at
10XAIAmplifier.com/diagnostic

In the next chapter, we'll flip the lens outward and look at your customer—how to use AI to personalize, connect, and scale service without losing soul.

But before that?

Get your team in the room. Run the warm-up. Because you don't automate your business without activating your people.

Chapter 9
Personalization at Scale

One of the biggest fears around AI is this: "If I automate, I'll lose the personal touch that makes us different."

Let's flip that fear.

Because done right, AI doesn't replace human connection—it **reclaims** it.

It gives you back time. It sharpens your message. It lets you show up consistently for more people, more personally, more often.

This chapter is about making your brand feel more human, not less—as you scale faster than ever before.

Why Personalization Feels Broken Today

Most businesses confuse personalization with plugging in a first name. That's not personal. That's templated.

What real personalization sounds like:

- "You understood my problem better than I could explain it."

- "That email felt like it was written just for me."
- "Your timing was perfect—I needed this today."

This kind of resonance used to require human touch. Now, with the right AI workflows, you can deliver it at scale—without burning out your team.

The 3 Layers of AI-Driven Personalization

To make this real, we break personalization down into three simple levels:

1. Smart Segmentation (Know Who You're Talking To)

Before you write or automate anything, ask:
- Who are we speaking to?
- What do they want most?
- What language do they use?

🧠 **AI Use:** Feed your customer survey responses or testimonials into ChatGPT and ask: "Group these into customer types and summarize their key desires and pain points."

💡 **Outcome:** Instant clarity on your audience personas—and how to speak to each one.

2. Tailored Messaging (Say It Like They Need to Hear It)

Great messaging doesn't just inform. It reflects, relates, and resonates.

🧠 **AI Use:** Use this prompt: "Write a 3-part email sequence for [customer type] who's stuck at [pain point], wants [outcome], and needs [specific nudge] to take action. Keep the tone warm and empathetic."

💡 **Outcome:** Copy that cuts through noise—and lands with heart.

3. Dynamic Delivery (Meet Them Where They Are)

Timing, channel, format—these are just as important as content. AI can help optimize all three.

🧠 **AI Use:** Use Zapier + AI tools to:

- Auto-tag leads by behavior or responses
- Trigger relevant content based on their actions
- Adjust email timing based on engagement patterns

💡 **Outcome:** More opens, more clicks, more conversion—with less lift.

Real-Life Personalization Use Cases

🎖 **Onboarding** Send a custom welcome video + AI-generated roadmap based on their quiz answers.

🎯 **Email Nurture** Use AI to group people by pain point and serve the right message at the right time.

📈 **Client Success** Analyze past client sessions and generate personalized recommendations for next steps.

💬 **Support & Retention** Auto-summarize support tickets, detect sentiment, and flag customers at risk of churn.

The Paradox of Scale

You've probably heard the phrase: "You can't scale intimacy." We disagree.

You **can't scale it manually**. But you **can** scale it with intelligent systems that amplify empathy.

Here's what that looks like:

- Consistent communication that adapts to the person, not just the list

- Timely responses that feel human, even if they're automated
- Messaging that's emotionally intelligent, not just technically correct

And when you pair automation with authenticity, you win trust—at scale.

Your Personalization Power-Up Plan

Here's how to implement this fast:

1. **Choose one audience segment.** Example: "Clients who bought in the last 90 days but haven't booked a call."
2. **Define one emotional driver.** Example: "They want results but feel stuck or unsure."
3. **Prompt AI for custom messaging.** Prompt: "Write a check-in email for someone who needs encouragement and next steps, tone: thoughtful but energizing."
4. **Automate the delivery.** Use Zapier or your CRM to trigger the message based on customer behavior.
5. **Track engagement and refine.** Let AI help you analyze what worked—and what to improve.

Another Fast Five AI Wins

Personalization strategies you can implement this week:

1. **"Create customer personas by analyzing your testimonials and reviews with AI."** → Instant audience clarity and messaging direction.
2. **"Set up behavior-triggered email sequences based on website activity."** → Timely, relevant communication that feels personal.
3. **"Use AI to customize onboarding based on how new customers found you."** → Immediate relevance and connection from day one.

4. **"Build dynamic email subject lines that adapt to engagement history."** → Higher open rates through personalized attention.
5. **"Implement sentiment analysis on customer support to flag at-risk accounts."** → Proactive retention instead of reactive damage control.

Scale Your Personal Touch

Personalization at scale isn't about having more data—it's about using intelligence to create more meaningful connections.

You don't have to choose between automation and authenticity. You can have both.

Our Strategic AI Readiness Diagnostic includes a personalization assessment that helps you:

☑ Identify where your customer journey feels impersonal
☑ Map opportunities for intelligent personalization
☑ Design automated workflows that maintain your brand voice
☑ Create systems that scale intimacy, not just efficiency

Based on your results, you may qualify for a complimentary 20-minute AI Strategy Call where we'll help you design personalization systems that make every customer feel seen and valued—without overwhelming your team.

Get your personalization diagnostic at
10XAIAmplifier.com/diagnostic

Remember This: AI Doesn't Diminish Emotion. It Delivers It More Often.

In the next chapter, we'll show you how to use AI not just as a tool, but as a compass for scaling your leadership and your legacy. You'll learn how to make smarter decisions, faster—and with deeper insight—by using AI as your thinking partner, not just your assistant.

But for now, start here:

- Pick one place in your customer journey that feels impersonal
- Plug in an AI workflow that adds warmth, timing, or relevance
- Test it, tweak it, and watch trust scale

Because in this new era, personalization isn't a feature. **It's a superpower.**

Chapter 10
Leading With Intelligence

Making Smarter Decisions, Faster—with an AI Thinking Partner by Your Side

You're not just building a smarter business. You're becoming a smarter leader.

The ultimate power of AI isn't that it can write content, summarize meetings, or automate tasks. It's that it can become a strategic co-pilot—a cognitive amplifier that helps you make better decisions in less time, with more confidence.

In this chapter, we'll explore how to move from using AI for efficiency... to using it for insight.

The AI Shift: From Assistant to Advisor

Most people use AI like a digital intern:

- "Write this blog post."
- "Summarize this email thread."
- "Fix my grammar."

But the most powerful leaders? They use AI like a partner:

- "What am I missing in this strategy?"
- "What's the counterpoint to this idea?"
- "What risks haven't I accounted for?"

That's the shift. From doing... to **thinking**. From taskmaster... to **strategist**.

Real-World Leadership Transformation: Brandon's Evolution

Six months after implementing AI systems at K to Z Interiors, something interesting happened to Brandon's leadership style. He wasn't just saving time—he was making fundamentally better decisions.

Before: Reactive Decision-Making

Brandon's typical day involved putting out fires. A client complaint would consume his morning. A supplier issue would derail his afternoon. Strategic planning happened "when things calmed down"—which they never did.

- His decision-making pattern:
- React to the loudest problem
- Make quick judgments based on limited information
- Second-guess decisions when new information emerged
- Feel overwhelmed by the complexity of running multiple business aspects

After: Intelligence-Augmented Leadership

Strategic Planning Example: When considering whether to expand into a new Louisiana market, Brandon's old approach would have been gut instinct plus maybe a few Google searches.

His new approach with AI:

1. **Market Analysis:** "Analyze the demographics, income levels, and housing market trends for Lafayette, Louisiana. What

opportunities and risks do you see for a window treatment business?"

2. **Competitive Assessment:** "Based on this research about local competitors, what would be our differentiation strategy and realistic market share potential?"
3. **Resource Planning:** "Given our current team of 6 installers and Kelsey managing operations, what would be the optimal expansion timeline to serve Lafayette without compromising Baton Rouge service?"
4. **Risk Mitigation:** "What are three potential scenarios for this expansion—best case, worst case, and most likely—and what contingencies should we plan for each?"

The Result: Instead of making an emotional decision based on "it feels right," Brandon made a data-informed choice with clear success metrics and backup plans.

Daily Decision Enhancement:

Brandon now starts each week with what he calls his "AI Strategy Session":

- Reviews key metrics with AI to identify patterns he might miss
- Uses AI to prioritize the week's most important decisions
- Tests major choices against different stakeholder perspectives
- Gets AI help translating complex decisions into clear team communications

Brandon's Reflection: *"I'm not smarter than I was six months ago, but my decisions are. It's like having a strategic advisor who never gets tired, never has a bad day, and always asks the questions I forget to ask."*

The Psychology of AI-Augmented Leadership

The most profound changes when leaders adopt AI aren't technical—they're psychological. Here's what we've observed:

1. From Analysis Paralysis to Confident Action

The Old Pattern: Many leaders get stuck in endless analysis, afraid of making the wrong decision with incomplete information.

The AI Shift: Leaders report feeling more confident making decisions faster because AI helps them quickly explore multiple angles and potential outcomes.

Real Example: A consultant we work with went from taking weeks to propose new service offerings to testing ideas within days. AI helps her rapidly model different pricing strategies, identify potential objections, and craft compelling positioning—turning strategic thinking into strategic action.

2. From Isolation to Augmented Perspective

The Lonely Leader Problem: Traditional leadership can be isolating. The higher you climb, the fewer people can give you honest, unbiased input on complex decisions.

The AI Solution: AI becomes an always-available thinking partner that can role-play different perspectives without politics, ego, or agenda.

Psychological Impact: Leaders report feeling less alone in decision-making and more confident that they've considered multiple viewpoints before acting.

3. From Perfectionism to Iteration

The Old Mindset: Many leaders feel pressure to get decisions "right" the first time, leading to delayed action and over-analysis.

The New Paradigm: AI makes it easier to test ideas, run scenarios, and iterate quickly. Leaders become more comfortable with "intelligent experimentation" rather than perfect predictions.

Case Study: A business owner struggling with pricing strategy used AI to model 15 different pricing approaches in one afternoon. Instead of agonizing over the "perfect" price, she tested three AI-recommended options with small customer segments and optimized based on real data.

4. From Reactive to Predictive

Traditional Pattern: Most leaders spend their time reacting to problems after they occur.

AI-Enhanced Pattern: Leaders begin using AI to identify potential issues before they become critical, shifting from firefighting to fire prevention.

Example: A growing agency owner now uses AI to analyze client communication patterns and project metrics. The system flags accounts showing early warning signs of dissatisfaction, allowing proactive intervention instead of reactive damage control.

Where AI Enhances Your Leadership

Here are the top five domains where AI enhances leadership clarity and confidence:

1. Decision-Making Under Pressure

Prompt: "List 3 pros and cons of this choice, from the perspective of a CFO, customer, and investor."
💡 **Use it to:** Map blind spots and see decisions from multiple angles.
Advanced Application: When Brandon had to decide whether to lay off a team member during a slow period, he used AI to explore the decision from the perspectives of: remaining team

members, the affected employee, customers who might experience service delays, and the business's long-term health. This multi-angle analysis revealed a creative solution: temporary hour reduction with cross-training opportunities.

2. Strategic Planning

Prompt: "Draft a 90-day launch plan for this new product, based on lean startup principles."

💡 **Use it to:** Structure priorities and timelines with proven frameworks.

Real Implementation: A service business owner used this approach to launch a new consulting package. AI helped break down the launch into testable phases, identify key metrics to track, and create contingency plans for different response levels.

3. Scenario Simulation

Prompt: "What happens if we delay our hiring by 60 days? What are the risks and benefits?"

💡 **Use it to:** Preview possible futures and reduce uncertainty.

4. Cross-Functional Communication

Prompt: "Translate this ops report into a clear executive summary for non-technical stakeholders."

💡 **Use it to:** Make information flow faster and clearer across departments.

5. Vision Amplification

Prompt: "Turn this bullet list of goals into a motivational narrative for our team offsite."

💡 **Use it to:** Align hearts and minds—not just deliver data.

The Confidence Multiplier Effect

One of the most significant psychological shifts we observe is what we call the "Confidence Multiplier Effect." Leaders who effectively integrate AI into their decision-making process report:

Increased Decision Speed: They make strategic choices 2-3x faster because AI helps them quickly process multiple perspectives and scenarios.

Reduced Decision Anxiety: The fear of "missing something important" decreases when AI helps systematically explore different angles.

Better Sleep: Several leaders have mentioned sleeping better after making AI-assisted decisions because they feel more confident they've done their due diligence.

Enhanced Team Trust: When leaders can explain their reasoning more thoroughly (thanks to AI analysis), team members report higher confidence in leadership decisions.

Intelligence Isn't Just Input. It's Framing.

One of the most underestimated leadership skills? **The ability to ask better questions.**

AI rewards the leaders who prompt with precision.

Compare: ✗ "What should I do about sales?" ✓ "Based on a 20% drop in inbound leads over 60 days, what three strategic actions could stabilize revenue within 30 days?"

The Psychology Behind Better Prompting: Learning to prompt AI effectively actually makes you a better strategic thinker. The discipline of framing clear, specific questions forces you to clarify your thinking and identify what information is truly relevant to your decisions.

Teach your team to think like this. Make it your new leadership language.

Your Thinking Partner Playbook

Here's how to integrate AI into your daily leadership rhythm:

Leadership Activity	AI Prompt Strategy
Daily Prioritization	"Based on these 10 tasks, rank by ROI and urgency. Recommend a top 3 focus."
Hiring Strategy	"Create a scorecard and ideal candidate brief for this role, including soft skills."
Revenue Review	"Based on last month's sales report, what trends, anomalies, or risks stand out?"
Board Prep	"Summarize these five updates into a 2-slide executive presentation."
Annual Planning	"Draft three potential 12-month roadmaps based on these five company goals."

What Smart Leadership Looks Like in the AI Era

You don't need to know every tool. You don't need to become a data scientist.

But you **do** need to:

- Ask sharper questions
- Design better systems
- Make decisions from insight, not overload

Leadership used to be about instinct and grit. Now, it's also about **leverage and discernment**.

From Gut Instinct to Augmented Insight

It's not either/or. You still lead from vision, values, and intuition.

But now you have:

- A **simulator** to test your strategy
- A **synthesizer** to compress complexity
- A **sparring partner** to challenge assumptions

You're not just leading with hustle anymore. **You're leading with intelligence.**

The Psychological Safety Net: Many leaders discover that having AI as a thinking partner actually frees them to trust their instincts more, not less. When they know they can quickly test gut feelings against data and multiple perspectives, they become more willing to take calculated risks and make bold moves.

Your 7-Day Leadership Intelligence Challenge

Here's how to put this into practice immediately:

1. **Choose one decision** you've been postponing.
2. **Frame it as a strategic prompt** for AI.
3. **Ask it to role-play** different stakeholder perspectives.
4. **Use the response** to clarify your next move.
5. **Repeat daily** for 7 days with a new decision each day.
6. **Reflect:** Where did AI help? Where did it challenge you?
7. **Share what worked** with your team to start building an "intelligence culture."

Another Fast Five AI Wins

Leadership intelligence strategies you can implement immediately:

1. **"Use AI to analyze your last three big decisions and identify your blind spots."** → Better self-awareness and decision-making patterns.

2. **"Create AI-powered scenario planning for your next strategic initiative."** → Reduced risk through better preparation.
3. **"Build executive summary templates that AI can populate from raw data."** → Faster, clearer communication to stakeholders.
4. **"Develop AI-assisted prioritization for your weekly leadership tasks."** → Focus on highest-impact activities consistently.
5. **"Use AI to translate complex strategies into compelling team narratives."** → Better alignment and buy-in from your organization.

Amplify Your Leadership Intelligence

The shift from reactive leadership to intelligent leadership isn't just about using AI tools—it's about fundamentally changing how you process information, make decisions, and communicate vision.

You don't have to navigate this transformation alone.

Our Strategic AI Readiness Diagnostic includes a leadership intelligence assessment that helps you:

- ☑ Identify your current decision-making patterns and blind spots
- ☑ Map opportunities where AI can enhance your strategic thinking
- ☑ Develop custom AI prompting strategies for your leadership style
- ☑ Create systems that amplify your best instincts with data-driven insights
- ☑ Build an "intelligence culture" within your organization

Based on your results, you may qualify for a complimentary 20-minute AI Strategy Call where we'll help you design leadership systems that combine your vision and values with the power of AI-enhanced decision-making.

Get your leadership intelligence diagnostic at
10XAIAmplifier.com/diagnostic

In the next chapter, we'll bring it all together with the Exponential Operating System—a new business playbook for multiplying results, reducing friction, and becoming the kind of organization that scales without losing soul.

But first, take a breath.

You're no longer guessing. **You're leading with clarity.** And you've got a new thinking partner to amplify your next move.

Chapter 11
The Exponential Operating System

Build a Business That Scales Without Burning Out or Breaking Down

You didn't come this far to work harder. You came to work smarter—*and* scale with purpose.

This chapter is your bridge from scattered AI experiments to a cohesive, scalable system. We call it your **Exponential Operating System**: a framework for amplifying every part of your business using intelligent workflows, empowered people, and leveraged leadership.

This isn't about using one tool. It's about designing a machine—with soul.

What Is an Exponential Operating System?

It's not software. It's not automation.

It's a way of running your business that aligns:

- Intelligence (AI, data, insights)
- Execution (systems, automations, workflows)
- Empowerment (team, leadership, culture)

- Impact (customer experience, growth, purpose)

At the heart of it? A strategy that's clear, simple, and scalable.

The Transformation Story: Marcus's Strategic Evolution

Marcus Rivera had built his business strategy consulting firm on one principle: deep, customized thinking for every client. But by late 2024, that commitment to excellence had become a prison.

The Breaking Point: Success Without Systems

Marcus's Reality:

- 3-person team (Marcus, senior analyst Jennifer, coordinator Alex)
- $480K annual revenue from 15-20 clients
- Every proposal took 8-12 hours to create from scratch
- Client reports required 15+ hours of manual analysis and writing
- Marcus worked 65 hours weekly, afraid to delegate his "strategic thinking"
- Opportunities lost because they couldn't respond to RFPs quickly enough

The Psychology Behind the Chaos: Marcus suffered from what we call "Genius Trap Syndrome"—believing that his unique insights couldn't be systematized without losing their value. He collected tools obsessively (14 different platforms for research, analysis, project management, and client communication) but never integrated them into a coherent system.

The Wake-Up Call: When a $150K client opportunity required a proposal in 48 hours, Marcus spent 36 hours straight creating it—and still lost the deal to a competitor who submitted earlier with a more polished presentation.

Building the Exponential OS: Marcus's 120-Day Journey

Phase 1: The Mindset Shift (Days 1-30) The hardest part wasn't technical—it was psychological. Marcus had to move from "I am the system" to "I design the systems."

Key Breakthrough: Realizing that systematizing his thinking didn't diminish its value—it amplified its reach.

Phase 2: Strategic Simplification (Days 31-60) Instead of 14 tools, Marcus built around 5 core systems:

- AI research and analysis (Claude + custom prompts)
- Automated client onboarding and project management
- Proposal and report generation system
- Client communication and relationship management
- Financial tracking and business intelligence

Phase 3: Team Empowerment (Days 61-90) Jennifer and Alex went from task-executors to strategic partners. Jennifer became the "AI Champion," helping optimize prompts and workflows. Alex evolved from coordinator to "Client Experience Designer."

Phase 4: Exponential Results (Days 91-120)

- Proposal creation time: 8-12 hours → 2-3 hours
- Report writing: 15+ hours → 4-6 hours
- Team capacity: 15-20 clients → 35+ clients
- Marcus's work week: 65 hours → 45 hours
- Annual revenue projection: $480K → $1.1M
- Client satisfaction scores: Up 40% (more consistent, faster delivery)

The Psychology of Systematic Transformation

Marcus's journey illustrates three crucial psychological shifts every leader must make:

1. **From Artisan to Architect Old Belief:** "My personal touch is what makes this valuable." **New Reality:** "My systematic thinking creates more value for more people."
2. **From Scarcity to Abundance Old Fear:** "If I systematize this, anyone can do it." **New Understanding:** "Systems free me to do higher-level thinking that truly can't be replicated."
3. **From Control to Leverage Old Pattern:** "I need to personally handle every strategic element." **New Approach:** "I design the frameworks that ensure strategic excellence at scale."

The 5 Core Layers of the Exponential OS

Think of this as the operating layers of a next-gen business:

1. Mindset Layer

From fear to focus. Your people (and you) understand AI not as a threat—but as an amplifier.

🗨 **Tools:** Training frameworks, resistance reframes, success story templates

Marcus's Implementation: Created "Strategic Thinking Amplified" workshops where the team learned to see AI as enhancing their expertise rather than replacing it.

2. Strategy Layer

From tactics to leverage. You don't chase every shiny tool. You choose based on friction points and ROI.

🗨 **Tools:** Diagnostic frameworks, workflow maps, priority filters

Real Application: Marcus mapped his entire client delivery process and identified that 70% of his time was spent on activities that could be AI-enhanced without losing strategic value.

3. Execution Layer

From chaos to clarity. Your daily ops are powered by smart systems—reducing manual work and wasted time.

🧠 **Tools:** Integrated automation platforms, AI writing assistants, smart CRMs

Case Example: Marcus's "Strategic Brief Generator" combines AI research, client questionnaire responses, and industry templates to create comprehensive project briefs in 30 minutes instead of 3 hours.

4. Culture Layer

From adoption to alignment. Your team feels supported, included, and confident in the changes.

🧠 **Tools:** AI champions program, learning pods, role-specific toolkits

Team Evolution: Jennifer went from research analyst to "Strategic Intelligence Designer." Alex evolved from admin coordinator to "Client Experience Architect." Both felt more valued and challenged in their enhanced roles.

5. Growth Layer

From one channel to exponential. You scale your message, service, and impact—without scaling burnout.

🧠 **Tools:** AI-powered marketing, personalization engines, evergreen systems

Exponential Results: Marcus's firm can now serve 3x more clients with the same team size, while delivering higher-quality, more consistent strategic insights.

The Psychology of Systems vs. Tools

The most common mistake we see is leaders who collect tools but never build systems. Here's the psychological difference:

Tool Collector Mindset:

- "This new AI tool will solve my problems"
- Reactive: responds to pain points with individual solutions
- Complexity increases with each addition
- Team confusion and tool fatigue
- Marginal improvements that don't compound

System Designer Mindset:

- "How can I design workflows that amplify our capabilities?"
- Proactive: creates integrated solutions that work together
- Complexity decreases as systems mature
- Team clarity and confidence
- Exponential improvements that compound over time

Marcus's Reflection: "I used to think I was being strategic by trying every new tool. Real strategy was stepping back and designing how all the pieces work together. The magic isn't in the individual tools—it's in the connections between them."

How to Build Yours in 30 Days

Week 1: Diagnostic + Alignment

- Map your current tech stack and workflows
- Identify biggest friction points and time drains
- Survey your team for AI concerns and ideas
- Define what "exponential" looks like for your business

Key Question: Where are you still the bottleneck, and what would change if you weren't?

Week 2: Simplify + Prioritize

- Choose one department or workflow to optimize first
- Pick 2-3 processes to automate or amplify
- Test AI-generated content or decision support
- Document what works and what doesn't

Success Metric: Achieve one measurable time savings of 3+ hours weekly

Week 3: Train + Test

- Run team AI training sessions (reference Chapter 8 exercise)
- Appoint 1-2 AI Champions to lead adoption
- Document wins, gaps, and team feedback
- Begin connecting individual tools into workflows

Cultural Indicator: Team members start suggesting their own AI applications

Week 4: Systematize + Scale

- Build your first "always-on" automated workflow
- Create regular review rhythms for AI system performance
- Plan next quarter's expansion areas
- Establish success metrics and optimization processes

Exponential Indicator: Systems begin improving themselves through usage data and feedback loops

The Resistance You'll Face (And How to Navigate It)

Every leader building an Exponential OS encounters predictable resistance:

Internal Resistance:

- **Perfectionism:** "The system isn't ready yet"
- **Control:** "I need to be involved in everything"

- **Identity:** "My value comes from doing, not designing"

Team Resistance:
- **Relevance:** "Will this make me unnecessary?"
- **Complexity:** "This seems like more work"
- **Trust:** "Can AI really handle our specialized needs?"

Client Resistance:
- **Authenticity:** "Is this still personalized?"
- **Quality:** "Will service standards drop?"
- **Relationship:** "Am I just talking to bots now?"

The Key: Address these concerns proactively with clear communication about how systems enhance rather than replace human value.

The Future Belongs to Builders

This isn't about becoming a tech company. It's about becoming an *exponential company*.

One that:
- Multiplies output without multiplying effort
- Keeps its heart while expanding its impact
- Leads with clarity instead of reacting with chaos

The exponential era won't wait. And it won't reward perfection. It rewards momentum, curiosity, and boldness.

Another Fast Five AI Wins

Exponential Operating System strategies you can implement immediately:

1. **"Map one complete workflow from start to finish and identify every manual handoff."** → Clear visibility on systematization opportunities.

2. **"Create AI-assisted templates for your three most common deliverables."** → Immediate time savings with consistent quality.
3. **"Build one automated sequence that connects two existing tools."** → Foundation for more complex system integration.
4. **"Design a weekly 'Systems Review' meeting to optimize AI workflows."** → Continuous improvement culture and momentum.
5. **"Train one team member to become your 'AI Champion' and system optimizer."** → Internal expertise and change management support.

Design Your Exponential Operating System

Building an Exponential OS isn't just about technology—it's about fundamentally redesigning how your business creates and delivers value. It requires both strategic thinking and practical execution.

You don't have to architect this transformation alone.

Our Strategic AI Readiness Diagnostic includes a comprehensive operating system assessment that helps you:

☑ Map your current workflows and identify integration opportunities
☑ Design custom automation sequences for your specific business model
☑ Create team adoption strategies that build excitement instead of resistance
☑ Develop systematic approaches to continuous improvement and optimization
☑ Build measurement frameworks to track exponential growth indicators

Based on your results, you may qualify for a complimentary 20-minute AI Strategy Session where we'll help you design a complete Exponential Operating System that amplifies your team's capabilities while simplifying your operations.

Get your exponential OS diagnostic at 10XAIAmplifier.com/diagnostic

In the final chapter, we'll bring it all together and issue your invitation: To become not just an AI user, but a *leader* in this new age of intelligent amplification.

Because your business isn't just meant to survive this shift. It's meant to **lead it**.

Chapter 12
Your 10X Invitation

This Isn't the End. It's Your New Operating Beginning.

If you've made it this far, you already know: This book wasn't just about tools, prompts, or tech stacks.

It was about *permission*. To think differently. To lead boldly. To build the kind of business—and life—that runs on clarity, not chaos.

The world isn't waiting for perfect timing. It's waiting for leaders who are brave enough to evolve faster than the fear.

That's you.

The Real Point of All This? Amplification.

You're not here to play catch-up.

You're here to 10X:

- Your time
- Your impact
- Your communication

- Your team's confidence
- Your ability to think, act, and grow at the speed of intelligence

That's what this book was designed to do: Not just show you *what* to do—but invite you into *who* you get to become.

The Transformation Is Already Happening

While you've been reading this book, thousands of business leaders have been quietly implementing the principles you've just learned. Here are glimpses of what's happening right now:

The Marketing Agency Owner

Who went from working 60-hour weeks to leading a team that produces 3x more content with AI-assisted creativity systems. She now spends her time on strategy and client relationships instead of fighting deadlines.

The Consultant

Who transformed his proposal process from 12 hours of manual work to 2 hours of strategic thinking supported by AI research and writing assistance. His close rate increased 40% because he could respond faster with higher-quality proposals.

The E-commerce Entrepreneur

Who built AI-powered customer service and inventory management systems that freed her from daily operational fire-fighting. She's now scaling into new markets while working fewer hours than when her business was half the size.

The Real Estate Team

That automated their lead nurturing, market analysis, and client communication workflows. They're now serving 50% more clients with the same team while delivering more personalized service than ever before.

What they all have in common: They didn't wait for permission. They didn't wait for perfection. They started with one system, proved it worked, then built from there.

From Tool-User to Thought Leader

Here's what separates those who dabble in AI... from those who dominate with it:

- They don't just react—they **design**
- They don't wait for certainty—they **act on clarity**
- They don't do it alone—they **build networks, systems, and partners**

That's where you go next.

The Psychology of Leadership in the AI Era

The leaders who thrive in this new landscape share three psychological traits:

1. Intelligent Impatience They're impatient with inefficiency but patient with learning. They want results fast but understand that mastery takes time.

2. Strategic Confidence They're confident in their vision but humble about their methods. They know where they're going but remain flexible about how they get there.

3. Collaborative Leverage They understand that the biggest wins come from amplifying others, not just themselves. They build systems that make their entire team more capable.

You're Not Behind. You're On the Edge.

Let's be real: The people around you might still be watching from the sidelines. Paralyzed by complexity. Stuck in resistance. Clinging to the way things used to work.

But you?

You're on the edge of a new era.

You've seen what's possible when mindset, message, and machine come together. You've started rewiring how you think—and what you build. You're not overwhelmed by AI anymore. **You're amplified by it.**

The Edge-Dweller Advantage

Being on the edge isn't just about adopting new technology first. It's about developing the mindset that allows you to:

- **See opportunities** others miss because they're focused on problems
- **Move faster** because you're comfortable with intelligent experimentation
- **Scale differently** because you understand leverage, not just effort
- **Lead confidently** because you're designing the future instead of reacting to it

This puts you in rare company. Most business owners are still trying to solve tomorrow's problems with yesterday's methods. You're building tomorrow's solutions today.

The Compound Effect of Early Action

Here's what most people don't realize about AI adoption: the benefits compound exponentially, but only for those who start early and build systematically.

Month 1: You save 5-10 hours per week with basic AI assistance
Month 3: Your team gains confidence and starts suggesting improvements
 Month 6: Your systems begin connecting and amplifying each other
Month 12: You're operating at a fundamentally different level— higher output, lower stress, better decisions

Month 24: Your business looks nothing like it did before, but it's unmistakably *you* at scale

The Psychology of Compound Gains: Early adopters don't just get a head start—they get a different trajectory. Each system you build makes the next one easier. Each workflow you optimize creates capacity for higher-level thinking. Each team member you empower becomes a force multiplier for further innovation.

Late adopters don't just start behind—they start on a steeper learning curve because the complexity gap has widened.

Your Call to Action: Step Into the 10X AI Amplifier

We built this platform not just as a book, but as a movement.

If you're ready to take the next step—personally or with your team—here's your pathway to exponential growth:

🔍 Start With Clarity: Take Your Diagnostic

Before you choose tools or strategies, get clear on where you are and where the biggest opportunities lie.

Our Strategic AI Readiness Diagnostic gives you:

- ☑ A comprehensive assessment of your current systems and gaps
- ☑ Personalized recommendations based on your business model and goals
- ☑ Priority rankings so you know where to focus first
- ☑ A roadmap that honors both your vision and your reality

Begin your transformation at 10XAIAmplifier.com/diagnostic

💬 Get Expert Guidance: Book Your Strategy Session

Based on your diagnostic results, you may qualify for a complimentary 20-minute AI Strategy Session with one of our Fractional AI Officers.

In this focused session, we'll help you:

- ☑ Clarify your highest-impact next moves
- ☑ Design a systematic approach to AI integration
- ☑ Address any concerns or resistance (your own or your team's)
- ☑ Create an implementation timeline that works with your current reality

This isn't a sales call—it's a strategic consultation designed to help you move forward with confidence and clarity.

💼 Access Your Implementation Tools

Whether you work with us or implement on your own, you'll have access to our complete toolkit:

- ☑ Templates and checklists for every chapter's strategies
- ☑ Ready-to-use prompts organized by business function
- ☑ Team training materials and resistance management guides
- ☑ Workflow automation blueprints and setup instructions

👥 Transform Your Team Culture

Use our proven Team AI Warm-Up exercise to build excitement and capability across your organization. When your entire team understands and embraces AI as an amplifier, your transformation accelerates exponentially.

The Three Paths Forward

As you close this book, you have three choices:

Path 1: The Observer

You'll think "That was interesting" and go back to business as usual. You'll watch others transform while staying stuck in familiar patterns. In 12 months, you'll wish you had started today.

Path 2: The Dabbler

You'll try a few tools sporadically, get frustrated when they don't immediately transform everything, and eventually give up. You'll miss the compound benefits because you never built systematic consistency.

Path 3: The Amplifier

You'll start with one system this week. You'll measure results, get feedback, and iterate. You'll build momentum through consistent experimentation and systematic improvement. In 12 months, your business will be unrecognizable—and unmistakably more *you*.

The choice is yours. But the opportunity won't wait.

The Future Isn't Written Yet. But You're Writing Yours Now.

AI will never replace you. But it *will* replace the version of you that doesn't adapt.

So this is your moment:

- To lead louder
- To scale smarter
- To amplify everything that makes you powerful

Not with hustle. With leverage. Not alone. With systems. Not someday. **Today.**

Your New Identity: The AI-Amplified Leader

You're not becoming less human by embracing AI. You're becoming more strategically human. More intentionally impactful. More systematically successful.

The AI-Amplified Leader:

- Makes decisions faster because they have better information

- Scales impact without sacrificing quality because they have intelligent systems
- Builds stronger relationships because they have more time for what matters
- Creates more value because they focus on what only humans can do
- Leads with confidence because they combine intuition with intelligence

This is who you're becoming. This is your new operating identity.

Welcome to your new advantage. **Welcome to the 10X AI Amplifier.**

Final Word
The Leader Who Moves

The most dangerous thing you could do after reading this book is say: *"That was interesting."*

This book wasn't written to entertain you. It was written to **activate** you.

Because here's the truth no one else will tell you: **AI will not wait for you.**

The opportunity won't freeze itself while you get ready. The marketplace is already shifting. Your competitors are already moving.

The good news? You're not behind. You're here. You've seen what's possible. You've got the mindset, the tools, and the frameworks.

The only thing missing now... is **motion**.

You don't have to master everything today. You don't need the "perfect" system. You simply need to start your next experiment.

- Take the diagnostic

- Test one prompt
- Automate one process
- Offload one task
- Build one small system

Let momentum do the rest.

Because the real advantage isn't the tech.

It's the leader who moves.

You're not becoming less human. You're becoming **amplified**.

This isn't the end of your AI journey. This is just the very first step in your exponential chapter.

Your future is waiting. Your team is ready. Your opportunity is now.

We'll see you on the front edge of what's next.

Take your Strategic AI Readiness Diagnostic at
10XAIAmplifier.com/diagnostic

The world needs leaders
who amplify intelligence with wisdom.
The world needs you.

Want to Go Beyond the Book?

The *10X AI Amplifier* isn't just a book. It's your invitation to transform how you lead, scale, and win in the age of intelligence.

If you're ready to accelerate what you've started here, **Samurai Partners** offers two powerful ways to go deeper:

📣 Invite Us to Speak

Co-authors Scott Sullivan and Jessica Fontana-Eddowes deliver transformational keynotes and workshops that help leadership teams unlock clarity, remove fear, and activate exponential growth using AI and mindset mastery.

📊 Schedule a 10X AI Readiness Assessment

This strategic diagnostic reveals where your business is leaking time, missing leverage, and ready for intelligent automation — with a roadmap tailored to your team, tech, and traction goals.

Ready to amplify?

👉 Visit SamuraiPartners.ai to book your experience.

About the Author – Scott Sullivan

Scott Sullivan is a **Leadership & AI Strategist**, 3X Amazon #1 Bestselling Author, and a trusted advisor to some of the world's most recognized brands.

Over the last two decades, Scott has helped leaders **scale influence, performance, and profit**—without sacrificing clarity or culture. As a Peak Performance Strategist with **Tony Robbins**, he delivered over **3,400 high-impact presentations worldwide**, training more than **100,000 leaders** at companies including **Mercedes-Benz, Microsoft, AT&T, MGM Grand, Caesars Palace,** and more.

He later **co-founded and scaled a startup from 4 people to 70 employees and 9-figure annual revenue in just 7 years**, earning contracts with enterprise clients like **Coca-Cola, UPS, and the Louisiana State Police**.

Scott's humanitarian leadership has earned him global recognition. He was invited to speak at the **United Nations** on *"One Person Can Make a Difference"*, awarded **International Peace Ambassador**, and honored in the **U.S. Congressional Record of Honor** for his Hurricane Katrina disaster relief efforts. His nonprofit has delivered **over 1 million food bags to hungry children**, and his story has been featured by *Good Morning America, People Magazine, Fox News, ABC, BBC*, and more.

Today, Scott is the **Co-Founder of Samurai Partners**, where he helps growth-minded leaders **leverage AI to unlock 15–30% gains in profit and productivity** through scalable systems and influence alignment.

✉ Learn more at **SamuraiPartners.com**

About the Author – Jessica Fontana-Eddowes

Jessica Fontana is a growth strategist, 3X Amazon #1 Bestselling Author, and Co-Founder of **Samurai Partners**. With 27+ years in digital marketing, she helps businesses implement **AI-powered marketing systems** that drive performance, profit, and scale.

Jessica's strategies have driven **20–40% gains in efficiency and revenue** for growth-minded brands across industries. Her work earned her back-to-back recognition from **Neil Patel** as one of the top eCommerce SEO minds in the U.S.

From managing **over 1 billion ad impressions at XOOM.com (NBCi)** in the early internet days to delivering award-winning results in today's AI-driven landscape, Jessica has helped brands evolve through **SEO, paid media, content ops, automation, and custom GPT solutions.**

She specializes in building **repeatable frameworks for acquisition, retention, and team productivity**—and has advised franchises, agencies, and operators scaling toward their next growth plateau.

At **Samurai Partners**, Jessica works alongside co-founder Scott Sullivan to align AI, systems, and leadership influence—so founders can scale smarter, faster, and without friction.

www.ingramcontent.com/pod-product-compliance
Lightning Source LLC
Chambersburg PA
CBHW060619210326
41520CB00010B/1402